Air and Weather

Full Option Science System
Developed at
The Lawrence Hall of Science,
University of California, Berkeley
Published and distributed by
Delta Education,
a member of the School Specialty Family

1487697
978-1-62571-441-1
Printing 8 — 3/2022
Standard Printing, Canton, OH

Table of Contents

What Is All around Us?

We can't see it, but it's all around.

It's in the sky.

It's in the treetops.

It's on the ground.

It's near and far, high and low.

What is it?

Air! We cannot see air, but we know it is there.
What happens when you blow up a balloon?
You fill it with air.
You can see that air takes up space.

You also can feel air
on your skin when the
wind blows.
Wind is moving air.

We can tell that air is there when we fly a kite.
The wind pushes against the kite and keeps it
in the sky.

We can tell that air is there when a parachute floats to the ground.
Air pushes up against the parachute so it comes down slowly.

Even this boat shows us that air is all around.
A propeller on the back of the boat pushes on the air.
The boat moves forward.

So, what is all around us, everywhere we go?
You know!

What Is the Weather Today?

Lots of things happen in the air.

The **temperature** might change from warm to cold.

Clouds might form, and **rain** might fall.

The wind might start to blow.

The condition of the air outdoors is called **weather**.

How do you know what the weather will be today?
One way to find out is to look outside.
If the sky is dark and cloudy, you know it might rain.
If there are no clouds and the **Sun** is shining, you
know it won't rain.
Well, it won't rain right away.

Clouds are made of little drops of water.
If there is a lot of water in a cloud, the cloud looks gray.
The water drops might get bigger and bigger.
They might fall as rain.

If the air is cold enough, the water drops might fall as **snow**.

Some days the clouds seem to sit on Earth's ground and water instead of floating in the sky. The air feels wet, and you can't see very far. These clouds near the ground and water are called fog.

The weather you see when you look outside
might change.
A day might start out bright and sunny.

Later, clouds might form.
Soon, the sky is filled with them.

If the clouds hold enough water and the drops
get big enough, it will rain.
A **storm** can blow in and out in an afternoon.
Or, a storm can stay around for days.

Weather is in the air.
Air is all around you.
You feel and see the weather
every day, all the time.
So, go outside.
Enjoy the weather.
It might change tomorrow!

Clouds

Clouds can be many sizes and shapes.
Watching clouds can help tell you what the
weather will be.

Some clouds are high in the sky.
They are thin and white.
These clouds are called cirrus clouds.
Cirrus clouds usually mean fair weather.

Some clouds are not so high.
They are big, white, and fluffy.
These clouds are called cumulus clouds.
Cumulus clouds can mean fair or stormy weather.

Some clouds are low to the ground.
They are gray and long.
These clouds are called stratus clouds.
Stratus clouds usually mean rain.

Can you find each type of cloud on the next page?

Water in the Air

Sometimes it rains.

Water falls from the clouds.

The water flows down sidewalks and streets.

It forms puddles in low spots.

Then, it stops raining.

The heat from the Sun warms the surface of the land.

The heat from the Sun also warms the air and water.

Soon, the puddles of water are gone.

The sidewalks and streets are dry.

Where does the water go?

The water dries up.

The water changes from a liquid to a **gas**.

Water as a gas is called **water vapor**.

The water vapor goes into the air.

We can't see water vapor in the air.
Water vapor is invisible.
Water vapor is all around us in the air.
The air moves from place to place as wind.

When the water vapor cools, it forms clouds.
Clouds are lots of liquid water drops.
Rain comes from the liquid water in clouds.

Rain falls from clouds as drops of water.
Precipitation is rain falling from clouds.

Changes in the Sky

When you look up at the sky, what do you see?
It depends on the time of day.
It depends on the time of year, too.

Sometimes you see the Sun.
The Sun is a **star** close to Earth.
You can feel the Sun's warmth and see it shine.
When you can see the Sun's light, it is daytime.

Where do you see the Sun in the morning?
It is low in the sky in the east.
Sunrise is in the east.

Where do you see the Sun just before it gets dark?
It is low in the sky in the west.
Sunset is in the west.

Where do you think
the Sun is at noon?

Where do you see the Sun at night?

You can't see the Sun because it isn't in the night sky.

The sky is dark without the Sun in the sky.

The Sun makes day and night.

When do you see other stars in the sky?
You see other stars only at night.
It has to be dark to see them.

Here are stars we see in the summer sky and
the winter sky.

Do they look like the same pattern of stars?

summer

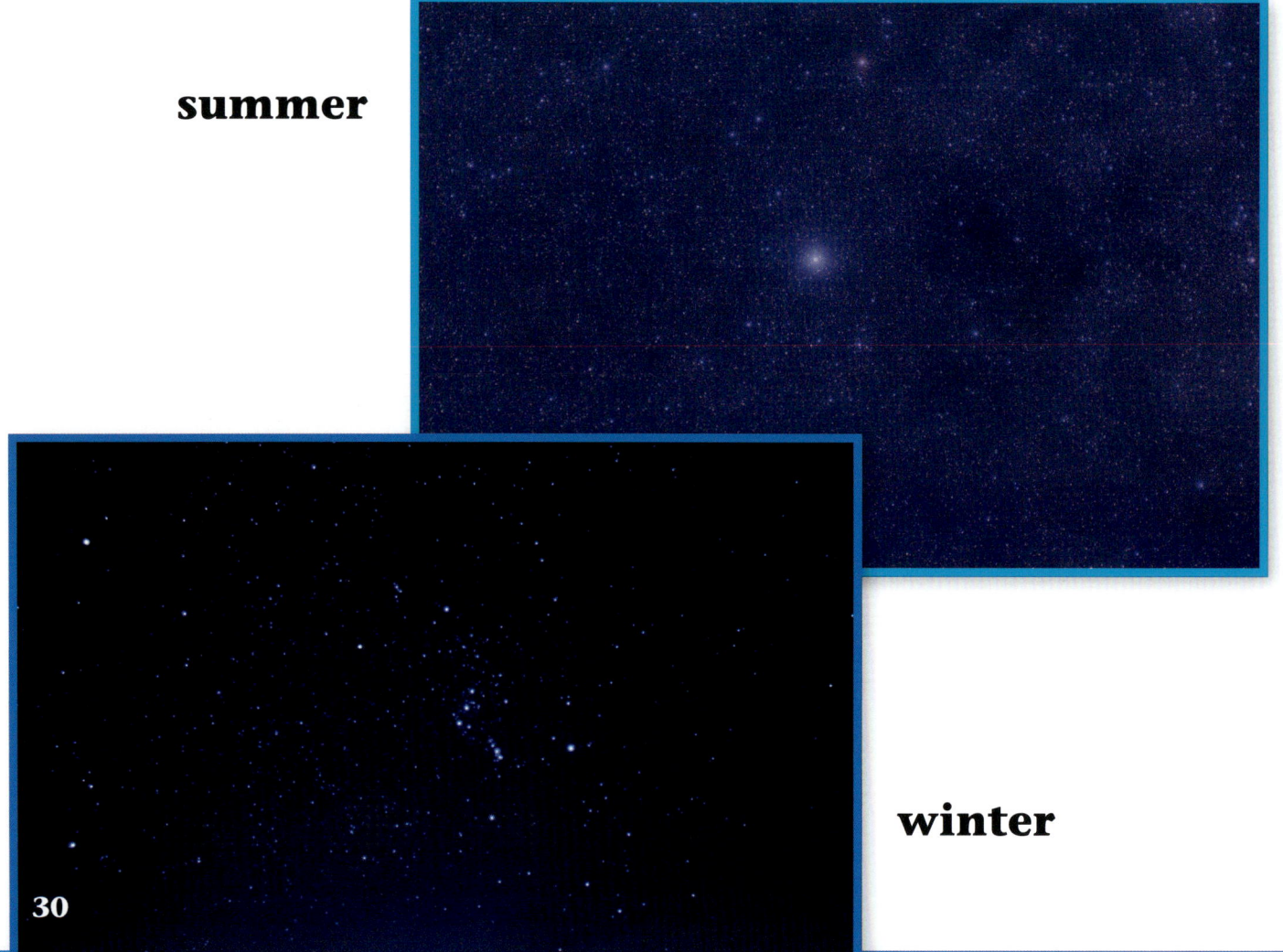

winter

Sometimes you see clouds in the sky.
It is easier to see clouds in the day sky.
But clouds can be in the night sky, too.

Clouds move with the wind.
They change all the time.
Sometimes clouds block the Sun.
They make **shadows** on the ground.

Sometimes you see the **Moon** in the sky.
You can see the Moon in the day sky
and the night sky.
But it looks different at different times.

This is a full Moon.
You can see a full Moon only at night.

Sometimes the Moon looks like a smile.
This shape is called a crescent Moon.
You can see a crescent Moon in the day sky
and the night sky.

Sometimes the Moon looks like a half circle.
This shape is called a quarter Moon.
You can see a quarter Moon in the day sky
and the night sky.

Sometimes the Moon looks like an egg.
This shape is called a gibbous Moon.
You can see a gibbous Moon in the day sky
and the night sky.

Look for the Moon every day or night for a month.
Record what you see and when you see it.
Is there a pattern to the Moon shapes you see?

Month _____

Sunday	Monday	Tuesday	Wednesday	Thursday	Friday	Saturday
1	2	3	4	5	6	7
8	9	10	11	12	13	14
15	16	17	18	19	20	21
22	23	24	25	26	27	28

Understanding the Weather

Some people study the weather.
They are called **meteorologists**.
They use instruments to gather
information about the weather.
Meteorologists **measure** the
temperature of the air.
They watch clouds form.
They measure wind speed and direction.

Weather balloons carry weather
instruments high into the sky.
The weather instruments gather
information.
This information helps meteorologists
tell us what the weather will be.

Sometimes weather is dangerous.
Meteorologists can help us know when to get
ready for a storm.

A **tornado** is a twirling, cloudy storm.
A tornado's winds blow around and around
very quickly.

A **hurricane** is a very strong, wet, and windy storm. Hurricanes form over warm ocean water.

A thunderstorm is a storm with lightning.
Lightning can be dangerous.
It is important to learn safety rules to be prepared for storms.

Thinking about Understanding the Weather

1. What is a meteorologist?

2. What does a meteorologist do?

3. Tell about different kinds of dangerous weather storms.

Resources

Things that people use are **resources**.
Resources from Earth are natural resources.
Resources that people change to make into
other things are human-made resources.

Plants and animals are natural resources.
Wood from trees is a natural resource.
People change wood from trees into lumber,
cardboard, and paper.
These things are human-made resources.

Cotton plants are a natural resource.
People use cotton to make thread.
The thread is used to make cotton
fabric for clothes.
Cotton clothes are good for
warm and sunny days.

Wheat plants are a natural resource.
Wheat seeds are harvested.
People use wheat to make cereal
and flour for bread and other foods.

Sheep are a natural resource.

People use the fur from sheep to make wool thread.

The thread is used to make wool fabric for clothes.

Wool clothes are good for cold and snowy days.

Water is a natural resource.
People use fresh water from rivers and lakes.
The water is treated to make it clean for
people to use.
It is used for drinking, washing, and cooking.

Earth materials like gravel, sand, and clay
are natural resources.
People use gravel and sand to make concrete
and asphalt.
People use clay to make bricks and pottery.

Air is a natural resource.
Plants and animals need air to live.

The Sun is a natural resource.
The Sun heats the air, land, and water.
The Sun makes our weather, too.

Look at the pictures.
Which things are natural resources?
Which things are resources made by people?

Seasons

In many areas, the **seasons**
bring different kinds of weather.

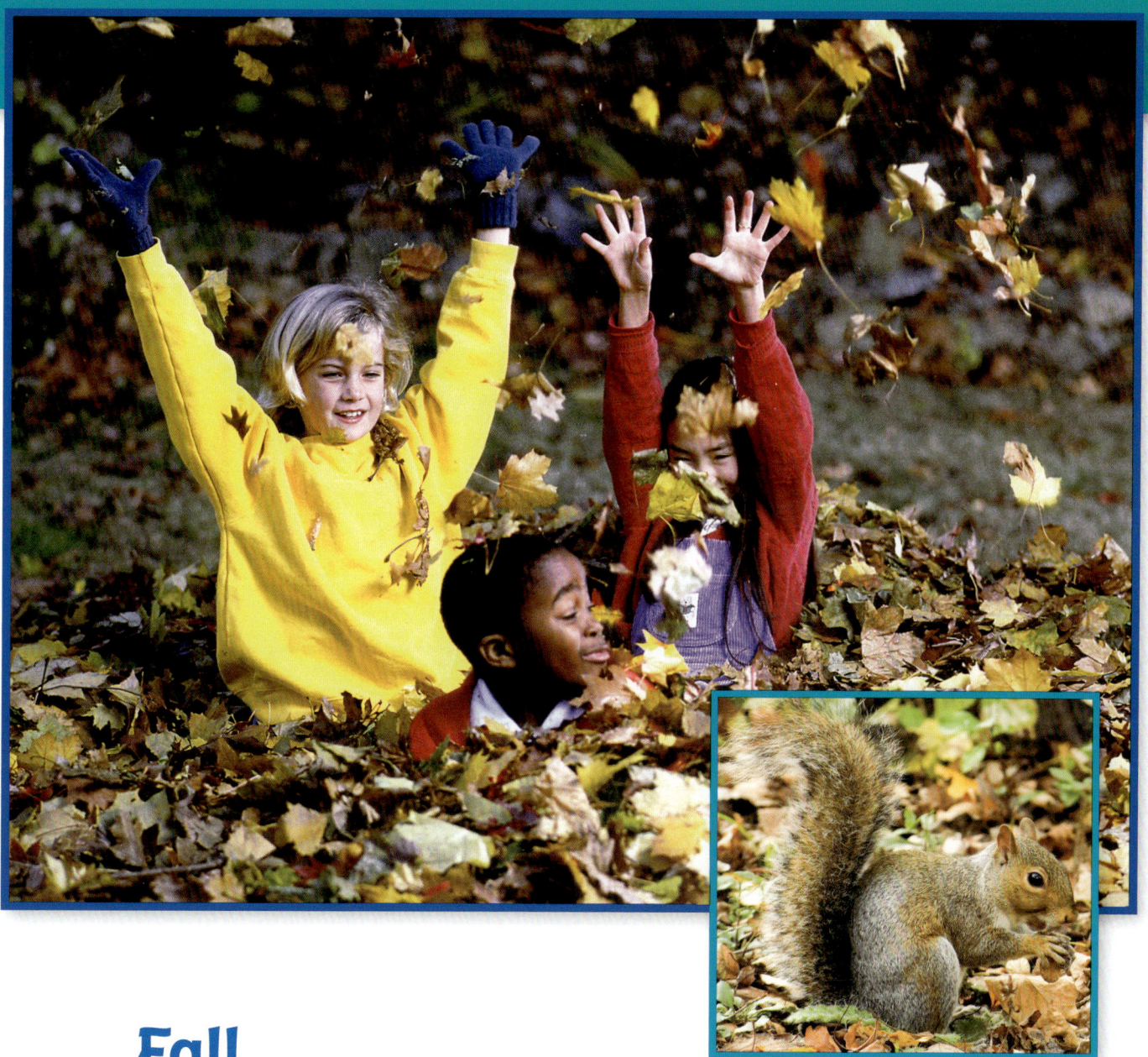

Fall

Leaves change color and drop from the trees.
Squirrels find seeds to eat.
A cool wind blows.
We put on our sweatshirts to play in the
leaves of fall.
The hours of daylight are decreasing.

Winter

Trees stand bare.
Few animals stir.
Snow falls to the ground.
We bundle up to keep ourselves warm
before we go sledding outdoors.
There are fewer hours of daylight
than in any other season.

Spring

Leaves grow on trees.

Flowers bloom.

Birds are building their nests.

The air warms up, and we go out to play

in the warm, soft breezes of spring.

The hours of daylight are increasing.

Summer

The Sun shines brightly on a hot summer day.
There isn't a cloud in the sky.
The trees give us shade.
We can make lemonade.
Then, we're off to the beach nearby.
There are more hours of daylight than
in any other season.

Knowing the weather in each season helps us make choices.

We can choose what to wear.

We can choose the kind of outdoor activity to do.

We even know how late we can stay out playing.

Thinking about Seasons

The seasons can make everything look different.
Which season do you see in each picture?
How can you tell?

Getting through the Winter

The weather affects many different plants and animals.
In spring and summer, the weather is warm.
Plants grow leaves.
Some plants grow flowers and fruit, too.
Animals can find lots of food to eat.

But in fall, the weather gets cooler.
Winter weather can be very cold in some places.
Some trees lose their leaves.
Many plants become **dormant** with no leaves,
flowers, or fruit.
Dormant plants are in a resting stage.

When plants are dormant, some animals
can't find food.
To get through the winter, some animals
become dormant, too.
This is called **hibernation**.

Bears and ground squirrels hibernate.
They make dens underground or in a hollow tree.
This is where they sleep until spring.

Other animals move for the winter.
They travel long distances to warmer places
where there is food to eat.
This travel is called **migration**.

What animals do you think migrate?

Many kinds of birds migrate.

Robins, ducks, and geese migrate.

Whooping cranes migrate, too.

In spring and summer, the cranes live in Canada.

This is where they lay eggs and raise their chicks.

When it gets cold, the cranes fly south for the winter.

Many kinds of insects migrate.

Dragonflies, ladybugs, and moths migrate.

Monarch butterflies migrate, too.

In spring and summer, the butterflies live in the northern United States.

This is where they lay their eggs.

The monarch caterpillars eat the plants that grow there.

In fall, many adult butterflies fly south, where it is warmer.

To get through the winter, some animals hibernate. Other animals migrate.

What animals in your area migrate?

Glossary

air a mixture of gases that we breathe **(4)**

cloud a group of very small water drops in the sky. Cirrus, cumulus, and stratus are kinds of clouds. **(9)**

dormant sleeping or not growing **(63)**

gas matter that can't be seen but is all around. Air is an example of a gas. **(22)**

hibernation when animals sleep through the winter **(64)**

hurricane a strong, wet, and windy storm that forms over warm ocean water **(41)**

measure to find the amount of something **(38)**

meteorologist a person who studies the weather **(38)**

migration when animals move when the season changes **(65)**

Moon the object we see in the night sky and sometimes during the day. Some of the Moon shapes we observe and describe are the full Moon, crescent Moon, quarter Moon, and gibbous Moon. **(33)**

precipitation rain, snow, or hail falling from the clouds **(25)**

rain one kind of weather that falls from the clouds as water drops **(9)**

resource a thing that people use. Natural resources, like the Sun or soil, are resources from Earth. Human-made resources, like cardboard or paper, are resources that people change from one thing into another. **(44)**

season one of four times of year that has different weather. Winter, spring, summer, and fall are seasons. **(54)**

shadow a dark area made by blocking the light from the Sun or other light source **(32)**

snow one kind of weather that happens when it is very cold. Frozen water falls from clouds. **(12)**

star an object in the sky that makes light and heat **(26)**

storm weather that has strong winds and can bring rain or snow **(16)**

Sun a star we see in the day sky. The Sun warms the land, air, and water. **(10)**

sunrise when the Sun rises in the morning **(27)**

sunset when the Sun sets (or goes down) in the evening **(28)**

temperature a description of how hot or cold something is **(9)**

tornado a twirling, cloudy, dangerous storm **(40)**

water vapor water as a gas **(22)**

weather the condition of the air outdoors **(9)**

weather balloon a balloon that carries weather instruments into the sky **(39)**

wind moving air **(4)**